The Weather by Heather

by Diana Perry

Illustrated by
Cody Sims

Copyright©2014 Diana Perry and Cody Sims

Published by
Biblio Publishing
The Educational Publisher Inc.
BiblioPublishing.com
BiblioBookstore.com
ISBN: 978-1-62249-287-9
Library of Congress Control Number: 2015952898

Acknowledgements

A very special thanks to Dr. Dave Walker who worked tirelessly to make sure all the weather facts were correct, and for hanging in there with me and for never giving up on this book!

> These mere words cannot express
> The way to say "Thank You!" best
> Kids everywhere will bond with Heather
> And learn many facts about the weather.
> Explaining weather was my mission to do;
> But this wouldn't have been, without wonderful you!

and to…..

Another very special thanks to the best illustrator in the world, Cody Sims, who didn't just illustrate this book – he brought the characters to life as his visual eye-candy illustrations made understanding weather a do-able learning process for kids of all ages. I can't wait to see the unique illustrations you will do with all my books. My success must be shared with you!

and to…..

Little Maddie H. for posing as the model for Heather. You are so pretty and smart and the perfect real life Heather. (Look for her image as 'Moira' the Hatter Fairy in 7 future books within 'The Periwinkle Fairy Adventures' series)

and to…..

Bob and Fran of Biblio Publishing, the fiction imprint of the decades old Zip Publishing, the Educational Publisher. You both gave me so much inspiration and guidance that I can't thank you enough.

and last, but not least…..

A huge gratitude to all you wonderful professionals who gave me blurbs for this book:

Lonnie Quinn, Chief Meteorologist on WCBS-TV NY who has also served as the National Weather Anchor for CBS' This Morning Saturday and CBS' The Early Show. He is also a 10-time Emmy Award Recipient. He is by far incomparable as being the nicest man on the planet earth, so important and famous but still found time to help me.

Dr. Dave Walker, Chief Meteorologist at WTRF Channel – 7, who formerly worked with NASA/NOAA and who possesses a wealth of knowledge on weather.

Executive Director, Sandy Whelchel of the National Writers Association, one of the most prestigious and respected writers associations; she has for decades tirelessly served both beginning and seasoned writers and authors across America. What would we do with you, Sandy?

…and to all the wonderful meteorologists not listed above throughout America who offered support and inspiration both to those who have already added their comments on *The Weather by Heather* to my website at DianaPerryBooks.com and those still contributing comments… thanks so very, very much, all of you!

To all the above and you out there reading this book….

Keep Looking Up……………there's weather out there!

CHAPTER ONE

"I'll never understand weather!" 8-year-old Heather exclaimed as she handed her list of weather topics to Miss Martin, the head librarian. "Third grade is really hard."

Miss Martin looked over the list. "Rain. Clouds. Snow. Well, there are a lot of topics in your homework assignment, but I know you can do this, Heather. Come with me." She led Heather to a corner table between the shelves. One by one, Miss Martin collected the books that Heather would need, stacking them into a tall pile in the center of the table. "Would you like me to help you?"

Heather sighed, "No, thank you. My teacher said we're big kids now and we need to try to do this by ourselves."

"All right," Miss Martin smiled as she started walking back to her desk. "If you get stuck, you know where to find me."

Heather glanced through the first three books and made some notes, but none of it made much sense to her. It wasn't long before she realized she needed help after all. Tearing off her notes, she tossed them into the trash can in the corner and, as she stood up to get Miss Martin, she noticed two of her classmates just opening the library door. "Oh, no. It's Julia and Karina." she said to herself. "I've got to get to Miss Martin before they do or she won't have time to help me." She leaped out of her chair and ran straight ahead, hoping she could beat them to Miss Martin when suddenly someone stepped out in her path. She had to grab onto the shelf to keep from slamming into the person just in front of her. "Ooohhh!"

PAGE 1

Quickly she turned left as she raced in between the first and second shelf, turning the corner at lightning speed without looking first. "Uumph!" She slammed hard into someone and fell backwards to the floor. "Ow! Ow!" Heather held up her arms trying to block the hard books that were raining down on her head.

"My goodness," A man said as he held out his hand to help her up. "Are you alright? What's your hurry anyway?"

Heather took his hand as he pulled her to her feet and glanced back toward Miss Martin. "Oh, no. It's too late. Julia and Karina got to Miss Martin first. Now I'll never get help with my science project."

"Well, since I got in your way, perhaps I should help you," the man offered. "I've been known to be very good at science."

Heather walked back toward her table, thinking that this man's voice sounded familiar. "Thanks, anyway, but my homework is on...." She recognized his voice as she turned around to face him. "...weather! It's you! You're Mr. Beedy! My parents watch you every day on TV."

He smiled back. "I am indeedy. What's your name?"

"I'm Heather." she answered as she led the way back to her corner table. "Here is my list of weather topics that I have to learn.

PAGE 2

Mr. Beedy looked over her list, then folded it and put it in his pocket. "Now, Heather, take good notes as we go and I'll try to make learning about weather fun. Let's start with fronts. A front is a bunch of air of the same temperature and is made up of tiny pieces called molecules. Just like all third graders stay together in the same room and all fourth graders stay in their room, molecules stay in their own 'classes' called fronts. Molecules come in two fronts; a warm front and a cool front and they never mix with each other. 1 Let's say this pile of books is a mountain. Put your arm out like mine and slowly come closer until our fingers touch. You be the warm front. I'll be the cool front." Slowly they moved their hands toward each other until their fingers touched.

"What happens when our fingers, I mean – fronts touch?" she asked.

He grabbed her wrist with his free hand, pulling it up over the books mountain. "Warm air molecules spread out having more air in between so they are lighter than cool air. Cool molecules huddle together to get warmer, just like we do when we're cold. They're packed in tightly and this makes the cool front get heavier. When the two fronts bump, the lighter warm front gets pushed up and rises up high over the mountain while the heavier cool front stays down low to the ground."

Heather made notes in her notebook as she repeated, "warm molecules are lighter so they rise up in the air and cool molecules are heavier so they stay low to the ground."

"I know someone who can help me teach you about weather," said Mr. Beedy as he led the way down the hall past the big round table where the Children's librarian was reading a story to smaller kids and into the room marked 'New Science Room'.

"You're Mr. Waite!" Heather called out to a man behind a tall desk with words on it. "I watch you on TV every Saturday."

"Why thank you," Mr. Waite answered. "This is where we film 'Waite For It'. Hi Dave."

"Good morning, Bill." Mr. Beedy said. "This is Heather. She has a homework assignment on the science of weather. I'm helping her to understand it and I thought your room would be the best place to show her many types of weather."

"Right you are," answered Mr. Waite. "Look around the room and see what you can use."

Mr. Beedy saw the helium balloons on the wall behind Mr. Waite and walked over to untie one. Walking back to Heather, he held it out to show her. "Ever wonder what makes balloons stay up? They're filled with Helium. Helium is a gas and gases tend to be warmer than air and what have we learned about cool air and warm air?"

"I remember." she answered. "Warm air is lighter so it rises."

"You're learning." Mr. Beedy said proudly. 2 "Let me show you air pressure. It's a big part of the weather. Hold out your hand, palm up. Air pressure is when you can 'feel' but not necessarily 'see' air." Pinching the end, he squeezed it tight allowing only a little bit of air to escape onto her palm.

"I feel the air pressure," Heather said, "even though I can't see it." Suddenly the balloon started making a squealing sound, getting louder and louder. "Make it stop, Mr. Beedy.

Mr. Beedy pinched it harder but that only made the squealing sound louder.

A voice was heard outside the door as the Children's librarian called out, "Who's making that noise?"

Mr. Beedy quickly let the rest of the air out of the balloon and set it aside, hoping no one would find out it was him. Taking a second balloon, he said, "We won't untie the end of this one. Hold still, Heather. I'm going to teach you about static electricity, which is found in storms. 3 He rubbed the balloon on her hair for several seconds, then lifted it up above her head.

Heather felt something funny and reached up to feel her hair standing straight up in the air. "What is happening, Mr. Beedy?"

"When I rubbed the balloon on your head," he explained, "static electricity was created making negative energy in the balloon and positive energy in your hair. When I pull the balloon away, the opposite energies reach out to touch each other. 4 Tear off a few tiny pieces of your paper please."

Heather ripped off several tiny pieces and put them on the desk top as Mr. Beedy held the balloon about an inch above the papers. The tiny papers seemed to jump up in the air and stick to the balloon. "Wow."

"One more static electricity experiment," Mr. Beedy offered as he walked over to the wall.

Heather followed.

5 He rubbed the balloon on the back of Heather's shirt for about a whole minute and then touched it to the wall. He pulled away his hand but the balloon stuck to the wall.

"I can't believe this." Heather gasped.

"Static electricity happens way up in the sky, too, when ice crystals bump into water droplets making flashing silver streaks in the sky. We call this lightning."

"What's ice crystals? she asked.

CHAPTER TWO

"Ice crystals," he said as he walked over to open the freezer door, "are tiny pieces of frozen water droplets way up in our atmosphere. The air is always freezing up there, even in the summer." 6 He scraped his finger against the side of the freezer and put the icy crystals into Heather's palm. "These are real ice crystals, just like you find up high in the sky; only these ones don't make weather."

"The ones way up high make lightning, right? Heather commented. "So what makes thunder?"

"The same ice crystals do." he answered. "They make a spark when they crash into each other causing the static electricity that comes down as lightning. The sound it makes comes down in a big boom that we call thunder. Would you like to 'feel' a little thunder, Heather?" He picked up a drum on a tall shelf and handed it to Heather, 7 "Hold this against your tummy and tell me what you feel." He patted the drum making soft drumming sounds.

"I can hear it but I don't feel anything." Heather said.

"Let's try again," Mr. Beedy said as he beat the drum harder making louder drumming sounds.

"I feel it," Heather cried out. "My tummy feels little tiny bumping things."

"Now who's making *that* noise?" the Children's librarian called out.

PAGE 6

"Let's talk about clouds; that's a lot quieter," Mr. Beedy suggested as Mr. Waite handed him a blue foam board with pictures of clouds. "Clouds are also made up of tiny drops of water and ice crystals. I'll start with this cloud," he pointed to the cloud on the left highest up on the board. "This is a cirrus cloud and it's made completely of ice crystals because, as I said, the air is below freezing way up there. Cirrus clouds look a lot like wispy white feathers and they warn us that rain is on the way."

He pointed to the next cloud down. "This cumulus cloud looks a lot like cotton or sometimes it can look like a big head of cauliflower. Cumulus clouds are the fun clouds with flat bottoms that take shapes in the sky. They can look like a face or a bear or any old thing. I always say this on TV – 'Fluffy white cumulus clouds and a sky of bright blue mean a day of nice weather for me and for you.'" Next he pointed to the light gray cloud below the cumulus cloud. "This dirty white stratus cloud looks like a big gray blanket across the sky."

"So which clouds make the rain?" Heather asked, jotting down the notes about clouds.

"Rain starts with the cloud called nimbus, a Latin word that means 'rainmaker'." He pointed to the cloud in the middle on the right. "It looks like a puff of smoke and it can block out the sun. Alone, nimbus clouds don't make rain but they carry a lot of 'precipitation' which means moisture in the air. When a nimbus cloud mixes with a stratus cloud," he pointed to the cloud below the nimbus cloud, "we call this a nimbostratus cloud. These clouds make a long steady rain shower."

"When a nimbus cloud mixes with a cumulus cloud..." he pointed to the very tall dark fluffy cloud with the dark bottom above the nimbus cloud, "that becomes a cumulonimbus cloud. These clouds are wild and unstable and can darken to a very dark blue or black. The air up there is swirling all about. Static electricity is formed; you felt that in your hair when I rubbed the balloon on your head. The fast moving air can start to spin; we call this a storm cell. This cloud gets the nickname of 'Monster Cloud'. Ice crystals crash into water droplets and we get silver streaks in the sky and loud noises. That, of course, is lightning and thunder. This cloud is also where hurricanes and tornadoes form. You can borrow this foam board to take outside anytime, Heather. 8 Compare these pictures with the clouds outside so you can see which clouds are in your sky. Now, this last cloud is fog." He pointed to the bottom cloud. "It's a type of stratus cloud that forms close to the ground early in the morning when the ground is cooler than the air above. It can also form over water and can be see-through or as thick as pea soup, as they say."

9 "How would you like to make your very own cloud?" Mr. Waite asked as he got out a tall clear cylinder, "and rain, too." He turned on a hot plate with a tea pot and in a few minutes the water inside it was steaming. "Dave why don't you take it from here..." he suggested as he poured the hot water into the cylinder filling it up.

"Okay I'll get the bag of ice," Mr. Beedy said. "I know where you keep that." He opened the freezer door once more and took out a baggie full of ice. After one minute he dumped out the water except for one inch. He struck 3 matches and held them over the top of the cylinder for one minute, then dropped them down inside. Quickly, he put the baggie with ice over the top. "Watch this," he said.

Heather bent in close to see a swirling white cloud appear off the ice and fall into the cylinder. Over and over, more white streaks of cloud appeared. Soon drops of water trickled down the sides. "Look, raindrops." She was so excited, she could hardly write down her notes.

Mr. Beedy grabbed a flower pot from the window sill with dirt but no plant and sat it down over the edge of the sink. Then he filled a glass with water and gently 'rained' over the flowerpot. 10 "A little rain is called a shower." The few raindrops were absorbed into the dirt, then he poured in much more water. 11 "Too much rain becomes a storm and when it storms too long and the ground can't absorb it all," he poured until the pot was filled with a pool of muddy water, 12 "it makes a pool of water on the surface. This is called a flood." It wasn't long before the dirt and water made a river of mud and spilled out into the sink. 13 "Too long of a flood can make a mudslide; that's when the flood acts like a river and travels picking up all sorts of trees, chairs, trash cans – anything in its' path and carries it along."

"Do you know what it's called when we don't get enough rain?" Mr. Beedy asked as Heather shook her head. "Give me a piece of your paper, please." 14 Heather tore off a piece of her notebook paper and handed it to him. He walked over to the sink and turned on the water, getting it all wet. Then he squeezed out as much water as he could and opened it back up. Walking over to the corner, he turned on the fan and held the paper up for the fan wind to dry it out. Then he held it out for Heather to touch. "This paper is all crinkled and cracked and very dry, just like the ground gets when it doesn't rain for a long time. The absence of rain is called drought."

"Here's more about ice crystals." Mr. Waite said as he brought out a black foam board with many different snowflakes. "Snowflakes always have six sides but no two are alike. Ice crystals up in the atmosphere are what snowflakes are made of."

"But," Mr. Beedy added, "ice crystals by themselves can't make a snowflake. They need something to stick to first, then they make six sides and get heavy enough to fall down as a snowflake. Want to know what they stick to?"

"Yes I do." Heather answered as she got ready to take more notes.

"There are basically four things," Mr. Beedy continued. "A tiny piece of meteorite from outer space can fall down into our atmosphere and ice crystals can stick to it and make a snowflake. It's so tiny you can only see it through a microscope so it's called a micro-meteorite. Also, from down on earth, there are three other things that get caught in an updraft and whisked way up in the sky where ice crystals attach to them and make snowflakes."

CHAPTER THREE

"What's an updraft?" Heather asked.

"It's a wind that blows straight up in the sky," Mr. Waite answered. "It looks just like this." He took a feather from under his desk and held it over his head. With a big breath, he blew on it to make it fly up a little way in the air. "Of course, this is a tiny updraft. The ones here on earth can pick up an impurity, that's something that isn't really part of ice crystals like the micro-meteorite, and take it way up high so the ice crystals can attach to it. Very little pieces of pollen, the yellow stuff you find in the centers of flowers, gets picked up in an updraft and taken way up to be made into snowflakes. Then there are tiny living things on the grass and plants called micro-organisms, also so tiny you can only see them through a microscope, and they get blown up to our upper atmosphere to be made into snowflakes. Tiny pieces of dust fly out of our buildings each time we open a door and they get caught in the updraft and come down in the center of snowflakes."

"Keep all this in mind as to what's inside a snowflake beside snow the next time you want to catch one on your tongue," laughed Mr. Beedy.

15 "You be the updraft this time." said Mr. Waite handing Heather the feather. "Keep blowing it back up each time it comes back down. Sometimes an updraft sends the snowflake back up in the upper atmosphere where more ice crystals attach to it, making it even heavier and then it falls down again. When this happens over and over again, so many ice crystals attach to the snowflake that it becomes just a big chunk of ice and when it's so heavy the updraft can't lift it anymore, it falls down one last time as hail. It's a big chunk of ice falling out of the sky and it can break windows and hurt people and damage anything it crashes down on."

"Why don't you show Heather how water is three different things, depending on the temperature?" said Mr. Waite. "You can use my tea pot of hot water. Just turn up the temperature to make…you know what."

"What a great idea," Mr. Beedy said as he put the tea pot back on the burner and turned it up high. "This should take about a minute to get hot. Let me tell you about the magic of water starting with water at room temperature." 16 He turned on the faucet and let the water flow. "Water in room temperature is liquid like you see out of this faucet." In a few minutes, the tea pot made a whistling sound and Mr. Beedy turned off the burner and pointed to the steam cloud coming out of the spout. 17 "When water is boiled, it becomes a gas called steam and looks just like a cloud." 18 Then he opened the freezer and took out an ice cube, placing it on Heather's hand, "but when water freezes it becomes a solid that we call ice." He took out five more ice cubes piling them up in her hand. "So what do you call a bunch of ice all stuck together?"

Heather shook her head, "I don't know but I wish you'd tell me quickly. My hand is freezing."

"Okay, then." Mr. Beedy chuckled, "It's an iceberg, of course. 19 Heather, hold your hand high up over the steam so you don't get burned. Do you feel your hand getting wet?"

Heather held her hand high over the steam and nodded.

"And just like with steam," he continued, "when tiny water droplets are up in the air like those that clouds can be made of, the air is filled with moisture. Moisture in the air is called precipitation."

Mr. Waite picked up his iced tea glass and moved it closer. 20 "Heather, wipe your fingers across this glass."

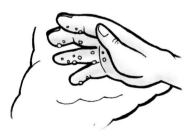 

Heather wiped her fingers across the glass and noticed this left 3 lines.

Mr. Waite continued, "When moisture sticks to an object it's called condensation."

As Heather added more notes, Mr. Beedy took out Heather's list from his pocket and looked it over. "You have winds on your list. Let's do those next." He walked over to the wall and picked up the floor fan. Holding the cord, he moved it closer to Heather, 21 "Heather, come stand right here in front of the fan. You be the earth and I'll be the winds; there are four in all." He turned it on medium and it blew back Heather's hair. "This is the north wind. In winter, this is the wind that brings in most of the cold weather, like the snow."

Next he moved to her right as the wind made her hair blow across her eyes. "This is the east wind. It comes from a great distance away from an area of high pressure to an area of low pressure. It can get to be fierce enough to make trees break, but usually it's just windy."

Then he moved the fan behind her, turning it on low. "This is south wind. This wind is gentle and warm and makes a lot of wind sounds. "Ooooo. Ooooo. People sometimes call this wind a warm summer breeze."

Heather closed her eyes and held out her arms, tilting her head back in this gentle wind.

"Hold on tight," he warned her as he moved to her left side. "For here comes my favorite wind, the west wind. It's the strongest which is why I like it the best." He turned the fan on high as it blew Heather's hair up and down wildly.

"This feels like a hurricane, Mr. Beedy." Heather complained as her hair blew wildly.

"Funny you should mention hurricanes," he said. "Let's do those next, and tornadoes, too."

 "I've got just the thing for a tornado," Mr. Waite said as he pulled out a long cylinder with a button to push. He filled it with water, then put one drop of green food coloring in it. "I think green works the best. You can really see the twister, which is a nickname for a tornado." He pressed the button and in a few seconds a green tornado was spinning in the cylinder.

22 "I can show you how to make one in a glass." Mr. Beedy offered as he took a tall clear glass from the counter and found a long spoon. Filling the glass with water, he added a drop of green food coloring. "Watch closely, for this only lasts a few seconds." He put in the spoon and stirred the water as fast as he could then quickly pulled out the spoon and in the glass was a spinning twister."

"Wow!" Heather exclaimed. "I can do this myself. What is the difference between a hurricane and a tornado and which way do they spin?"

 "Good questions," Mr. Beedy said. "First of all, a tornado forms over land and is more narrow and a hurricane forms over the sea and looks like a very wide spinning circle of storm. Now for your second question - Bill, help me out; 23 you spin clock-wise and I'll spin counter-clockwise." They both began to spin as Mr. Beedy continued, "Heather, clockwise is the way the hands on the clock turn and counter-clockwise is the opposite direction. Now you have to guess which way is the correct way that a hurricane or tornado spins."

Heather watched as the two men began to get dizzy but she couldn't figure it out.

"You better tell her, Dave, before I fall over." Mr. Waite said as he stopped spinning.

"I'd rather show her," he said taking a pinwheel from the shelf and handing it to Heather.

"Heather, blow on this side of the pinwheel and tell me which way it spins."

Heather took a deep breath and blew. "It went clockwise. So that's the answer."

"Come stand on my other side and try that again." Mr. Beedy said.

Heather crossed over to his other side and blew on the pinwheel again. "I'm confused. This time it turned counter-clockwise. So what is the right answer?"

"It's both!" Mr. Beedy exclaimed. "Most hurricanes and tornadoes in the Northern Hemisphere, the top half of the planet, spin counter-clockwise but they have been known to spin clockwise, too. In the Southern Hemisphere, they usually spin clockwise, but again they have been known to spin counter-clockwise."

CHAPTER FOUR

"To be more specific," Mr. Beedy continued, "a tornado is a funnel-shaped cumulonimbus cloud that can be up to one mile across and can spin more than 300 miles per hour, is born over the land and is made from a single thunderstorm. It can travel up to 50 miles and can be up to a mile across. It can last anywhere from 10 minutes up to one hour. We have a section of America known as Tornado Alley made up of Arkansas, Iowa, Kansas, Louisiana, Minnesota, Nebraska, North Dakota, Ohio, Oklahoma, South Dakota and Texas. The United States is struck by thousands of tornadoes each year. Tornadoes can form anytime but the season is March through August.

"A hurricane is a dangerous tropical storm with heavy rain and can average up to 100 miles across but can get even wider. It starts off as a Tropical Depression where the spinning winds are 38 miles per hour or less, not too dangerous. If that builds up to be 39 to 73 miles per hour then it becomes a Tropical Storm. Once it gets up to 74 miles per hour, we call it a hurricane. It is born over the ocean and is made up of several to dozens of storms. It can travel several hundreds of miles and last from 1 to 31 days, but starts to die out once it reaches land. Larger hurricanes have been known to go over 100 miles inland before they die out. When the spinning funnel storm is over the Atlantic Ocean or the eastern Pacific Ocean, we call it a hurricane. In India and the South Pacific Ocean, they call it a Cyclone and in the western North Pacific Ocean and the Philippines, it's called a Typhoon. Over Australia, they call it a Willie-Willie. A hurricane can make tornadoes come out of it when it travels on land, but a tornado can't make a hurricane. Another way a hurricane is different from a tornado is that inside the middle of a hurricane is a calm piece of air called 'the Eye' which can be 12 to 30 miles across. Only about 6 to 7 hurricanes normally strike or come close to the United States each year. Hurricanes can form anytime but the season is June through November. Both hurricanes and tornado are rated from the mildest to the most dangerous.

"The formula to make a spinning wind is this: 2 winds coming together from opposite directions, cool air from a cool front, warm air from a warm front and this starts it all spinning. It looks like a tornado up in the sky and we call this a funnel cloud. One force, gravity, reaches up and pulls it down to earth. To show you how gravity works, Heather, take this spoon and drop it."

24 Heather held out the spoon and let go. It dropped to the floor.

"That was gravity that pulled it down from the air." he explained, "but there are two other forces that keep a tornado or hurricane spinning. One is a force that pulls on the spinner. It's called Centripetal Force. I'll show you how it works." **25** He picked up a glass from another shelf that had a large marble in it. "Now this only works if the middle of the glass if wider than the top." He put the marble on the desk and the glass down over it. He began moving the glass around in fast circles making the marble go round and round inside. Soon it was pulled up into the middle of the glass. Then Mr. Beedy pulled the glass up in the air and the marble didn't fall out. "This is the pulling force that keeps it spinning; it's called Centripetal Force."

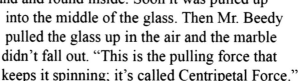

"Cen-tri-pet-al Force," Heather repeated as she wrote it down.

26 Mr. Waite pulled out a bowl and filled it half way with water and handed it to Mr. Beedy along with something that looked like a straw arrow. "This is just a straw that I bent in 3 equal sides. I stuck this skewer stick in the middle of the bottom triangle part and taped the two ends against the sharp end of the stick. Then I half way cut the two triangle points where the straw bends."

Mr. Beedy took over, "I'm going to use Centrifugal Force, the pushing force that keeps it spinning, to push the water up against gravity and come out the cut ends of the straw." He put the sharp end of the stick in the water and, holding the flat end with his thumb and forefinger, he spun it left and right as fast as he could and water poured out of both ends of the straw."

"Cen-tri-fu-gal," Heather said as she wrote that down in her notebook.

27 "Let me show you my experiment on how the earth's core looks," Mr. Waite said, as he pulled out a big candle which he lit. After a few minutes, the candle started to melt. "The inside of our earth is made of hard rock, and the center, or core, is so hot that it makes the rock melt and flow just like this candle wax does."

"Also those rocks come in huge sections called Tectonic Plates and they're always moving and pushing against each other," 28 Mr. Beedy added as he grabbed 2 very large books from Mr. Waite's book shelf pushing them together. Soon, one book pushed up over the other. "When one of them breaks up over the other, this causes the ground above to shake and sometimes break and crack. This we call an earthquake. Stand on this small rug here and hold on tight to the counter."

29 Heather stood on the rug and held on very tight as Mr. Beedy grabbed the rug on one end and tugged it just a little making Heather's leg jerk. "This is what a tiny earthquake feels like," he explained. "Hop off the rug so I can show you the next experiment."

30 Heather hopped off and he moved the rug up and down in slapping motions making it look like waves. "But, when this happens under the ocean. This causes a wave to come up called a t-s-u-n-a-m-i, pronounced sue-nom-ee. This wave can be only 1 foot tall or as big as 30 feet tall; however sometimes it becomes a giant tsunami wave and can get up to 1,720 feet tall. It can be from 98 feet wide up to 1,722 feet wide and can travel as fast as 500 miles per hour.

"Also when these tectonic plates move and break and are too close to a volcano, the molten rocks can escape up through a crack to come out the top of the volcano. We call this molten rock lava. Volcanoes erupt in bright yellow-orange-red colors. Let's make one now." Mr. Beedy suggested.

31 Mr. Waite brought out a round aluminum pie pan, a spoon, a glass vase larger at the bottom, red and yellow food coloring, yellow dishwashing liquid, an opened box of orange Jello, baking soda, vinegar, a bottle of water and a measuring cup and nodded to Mr. Beedy.

Mr. Beedy put the vase in the center of the pie pan. Next he measured ¼ cup of vinegar and poured it inside the vase. Then he measured 1 cup of water and poured that in. "Pay attention, Heather. This is my special recipe for a very erupting volcano."

"Wow!" Heather exclaimed.

Then Mr. Beedy poured in just enough water to cover the bottom of the measuring cup. To that he added 1 spoonful of the Jello, 2 heaping spoons of baking soda and 1 squirt of the dish liquid. "I use yellow dish liquid because I want my lava to come out orange. Also, the Jello isn't necessary for our volcano to erupt but it makes the lava look more real, not just watery." Then he added 2 drops of yellow food coloring and 2 drops of red and stirred this all up. "This brings me to the science of chemistry. Chemistry is the science that brings all kinds of other sciences together. The baking soda and the vinegar combine to make a 'chemical' reaction which makes carbon dioxide and it looks much like lava. Everything else works with this mixture to make our orange volcanic eruption." He handed the measuring cup to Heather, "Go ahead, you pour in our volcano formula."

Heather poured it in and was barely able to get her hand out of the way when a rush of orange lava erupted out of their volcano." Heather giggled and laughed.

Mr. Beedy took out Heather's list from his suit pocket. "Looks like we've covered everything on your list…except eclipses."

"Yay! It's time for our eclipse show," Mr. Waite said as he walked over to the doorway. "I'll get the room light when you're ready, Dave."

Mr. Beedy nodded in agreement and led Heather over toward the doorway where a table with a lamp stood. "An eclipse happens when the earth, which is orbiting around the sun… and the moon, which is orbiting around the earth, all line up together. There are two kinds; lunar and solar. Let's start with solar eclipses.

"The word 'solar' is a Latin word that means sun. 32 These eclipses happen in the day." He turned on the lamp and removed the lamp shade. "This light bulb will be our sun. You be the earth and stand in the middle." He moved Heather closer to the lamp. Then he pressed the button on Mr. Waite's computer and out popped a round disc. "We can pretend this is our moon. Watch what happens when our moon gets in front of our sun. All three of you are in a straight line." He slowly passed the disc in front of the lightbulb and most of the light was cut off. "You can see the bright glow around the edge of our moon as it blocks part of the sun's light from reaching the earth. The sky slowly gets dark as a shadow is cast over the earth as you can see on your face, Heather. This is what it looks like in the sky during a solar eclipse.

"Now for a lunar eclipse." 33 He turned off the ceiling light so that the lamp was the only light in the room and held the disc-moon behind Heather. "A lunar eclipse happens when the earth moves between the Sun and the moon, blocking part of the Sun's light from reaching the moon. During a lunar eclipse, you'll see the earth's shadow on the moon."

"We had to make special boxes in second grade last year, so we could look through them to see an eclipse." Heather told him.

"Your teacher was smart." he told her. "Looking directly at the sun or at an eclipse can damage your eyes. But you can look at our eclipses here. They're only pretend." He put back the disc and put the lamp shade back on as Heather turned on the ceiling light.

"Weather science is actually fun. So, if I want to be like you when I grow up, I'll have to learn about chemistry?" Heather asked.

"You'll have to learn many sciences." he answered.

The study of what makes the earth's core melt and how the hot lava escapes up out of the volcano is called volcanology.

The study of the materials that make up the earth, such as rocks, gems, minerals and crystals, and the conditions that form them is called geology

When we study the earth's features, like mountains, oceans and valleys, that's called geography

The study of ice and icebergs is called glaciology

The one science that explains how all the other sciences work and what everything is made of is, of course, chemistry

When we study about earthquakes and tsnuamis, that's called seismology

"Show me more." Heather was enjoying learning about weather.
"Sorry I have to go, now. It's time to report the weather on TV." he said as he walked away, "I'm going to say a special hello to you on the air tonight."

She finished her report and got an a-plus
and even showed her class a cloud that was cumulus.
Fifteen long years have passed since then then.
Do you wonder how little Heather has been?
Well turn on your TV when it's time for the weather,
for it's there you will find her on.....

…..The Weather by Heather

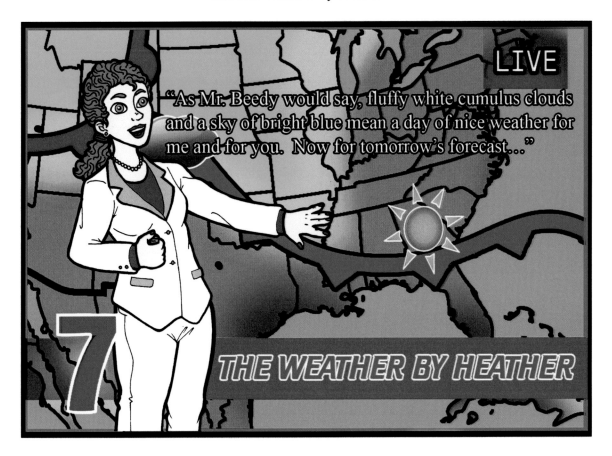

PAGE 21

Weather Fun Facts

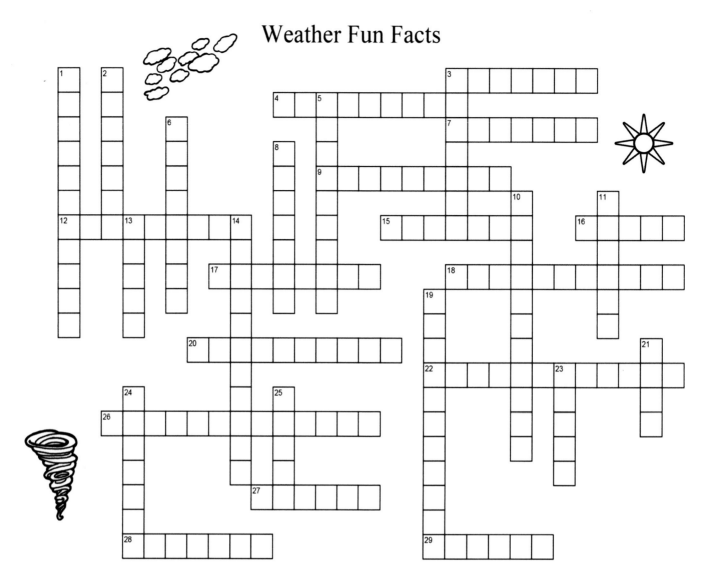

ACROSS

3 the loud boom we hear when ice crystals crash together in a storm
4 tiny pieces of air
7 a wind that picks up an impurity taking it way up in the upper atmosphere to be made into a snowflake
9 a spinning wind born over the ocean
12 each one has 6 sides but no 2 are alike
15 one of these happens when the earth, moon and sun are in a straight line
16 what we call when we get too much rain that makes a pool on the ground
17 name for the huge rocks way down below the ground
18 the invisible pushing force of a hurricane or tornado
20 when this happens, the ground shakes and cracks open
22 moisture from the air on an object
26 moisture in the air
27 groups of air of the same temperature
28 a spinning wind born over land
29 what we call a light rain

DOWN

1 (2 words) you can feel this even when you can't see it
2 the tall mountain that erupts lava
3 one of these can form on the ocean surface when an earthquake occurs on the ocean floor
5 the big silver streaks in the sky when ice crystals crash together in a storm
6 name for a really bad flood that travels like a river picking up all sorts of stuff
8 the invisible force that pulls down a funnel cloud and keeps everything held to the earth
10 the invisible pulling force of a hurricane or tornado
11 some of these are called cumulus or stratus
13 4 in all, the direction they're from gives them their names
14 the static stuff made by ice crystals in a storm; this is what we call lightning
19 (2 words) tiny frozen water droplets found both in our upper atmosphere and in the freezer
21 name for the hot center of the earth
23 name for a heavier rain
24 the absence of rain
25 the magic stuff that can be a liquid, gas or solid

TYPES OF WEATHER WORD SEARCH

```
H Z P Q R T E K A U Q H T R A E L H
R Y R C E N T R I P E T A L M M D Z
Y P E O M H H C M B R H R F C L K V
C X C N R M G L F K C D Q B T W F D
O R I A Z R U L T C H R N K Z G D K
N K P C K H O R M S E L U C E L O M
D X I L Z O R L A G U F I R T N E C
E Y T O D N D N S R M L N J C Y Z K
N G A V X M K C Y D L M W W K W S B
S S T O R M Y Y K K N K P C R E D H
A J I O R L I G H T N I N G S Y U U
T R O C R T I M H Z U M W P V T O R
I D N M N N R M P T W P I D H V L R
O R K O H Q A L A T N L D U R D C I
N T R T W K K D G N C W N R R D J C
C F V V F Z B L O E U D R N A Z V A
K V K T B N Z V F Q E S Q Q F F Q N
L R F C Q H F R N R K J T V L Y T E
```

CENTRIFUGAL
CENTRIPETAL
CLOUDS
CONDENSATION
DROUGHT
EARTHQUAKE
ECLIPSE
FLOOD
FRONT
HURRICANE

LIGHTNING
MOLECULES
PRECIPITATION
STORM
THUNDER
TORNADO
TSUNAMI
UPDRAFT
VOLCANO
WINDS

COUNTING GAME

Count the number of snowflakes.
Count the number of raindrops.
Then count the number of rainbows.
Finally, add them all together; what's the total number of them all?
Check your answers on the answer page.

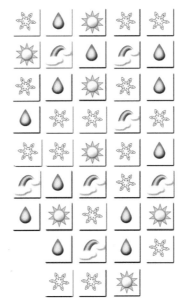

HIDDEN ITEM'S SEARCH

On what number page is each of these items found? Check back through the book and look for each item, then write the page number on the blank.

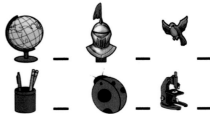

CODE MODE

Fill in the squares below with the letter that is found above the number.

If you do it correctly, you should have a sentence that tells you a fun weather term.

A B C D E F G H I J K L M N O P Q R S T U V W X Y Z
26 25 24 23 22 21 20 19 18 17 16 15 14 13 12 11 10 9 8 7 6 5 4 3 2 1

☐ ☐ ☐   ☐ ☐ ☐ ☐ ☐ ☐ ☐ ☐ ☐   ☐ ☐ ☐ ☐
26 15 15   8 13 12 4 21 15 26 16 22 8   19 26 5 22

☐ ☐ ☐   ☐ ☐ ☐ ☐ ☐
8 18 3   8 18 23 22 8

MATCH BATCH

Below are 6 types of weather. Can you draw a line from one on the left to the one that matches on the right?

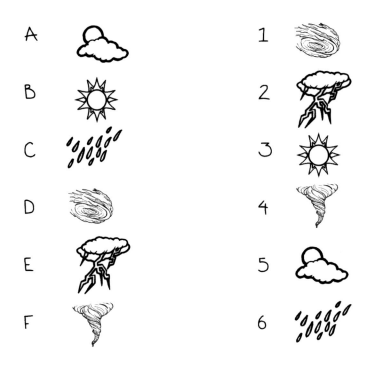

TORNADO MAZE

A tornado is on the way to Heather's town. Can you draw a line to lead the tornado to the open field and away from all the houses?

RHYMING GAME

Draw a line from the word on the left to the one that rhymes with it on the right.

TORNADO	GEOLOGY
HEATHER	FREEZE
BEEDY	MUD
FLOOD	WEATHER
GLACIOLOGY	INDEEDY
BREEZE	VOLCANO

SPELL TELL

There is a letter missing from the each of the words below. Can you tell which one? Fill in the blanks with the right letter so each word will be spelled right. The missing letter are listed below. Cross off each one as you use it. Use a pencil so you can erase if you get it wrong. To help you out, you can find every word spelled correctly in the book.

_SUNAMI	TSUNA_I	FLOO_	TOR_ADO
GRAVIT_	HUR_ICANE	RA_N	SNOWF_AKE
LIG_TNING	MOLE_ULES	_IND	EARTH_UAKE
_OLCANO	ECLI_SE	CUMUL_S	SUNLI_HT
IC_BERG	BLIZ_ARD	ECLIP_E	DR_UGHT

A C D E G H I L M N O P Q R S T U V W Y Z

Color Heather and Mr. Beedy

Continue for Answers

ANSWERS

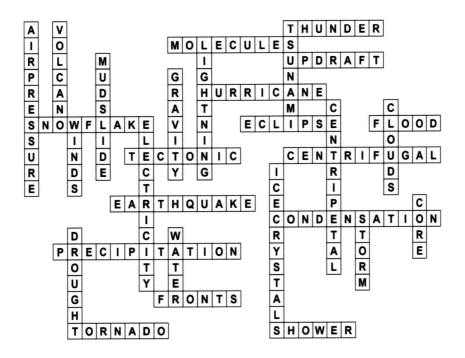

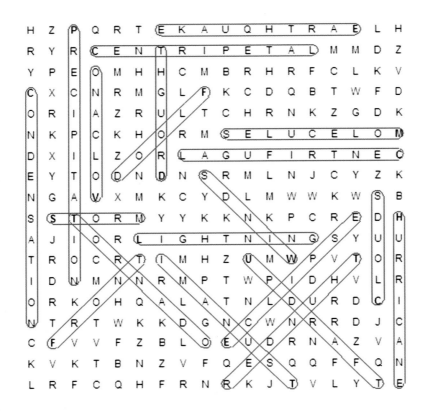

COUNTING GAME: 16 Snowflakes, 12 Raindrops, 7 Rainbows, 7 Suns   Total: 42

CODE MODE:

ALL SNOWFLAKES HAVE SIX SIDES

HIDDEN ITEMS SEARCH

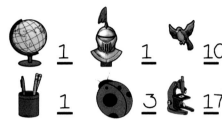

1   1   10
1   3   17

MATCH BATCH:

A-5, B-3, C-6, D-1, E-2, F-4

TORNADO MAZE

RHYMING GAME

Tornado.....Volcano
Heather...Weather
Geology...Glaciology
Freeze...Breeze
Flood...Mud
Beedy...Indeedy

SPELL TELL

TSUNAMI   TORNADO   HURRICANE   SNOWFLAKE   LIGHTNING
RAIN   MOLECULES   WIND   EARTHQUAKE   VOLCANO
ECLIPSE   CUMULUS   SUNLIGHT   BLIZZARD   GRAVITY
FLOOD   THUNDER